Tornados

A Layman's View

by
Jerry Figgins

ISBN-13: 978-1466339217

ISBN-10: 1466339217

Mail comments to:

Jerry Figgins   (717) 464-1741   jerryfiggins@yahoo.com
Apt. A-404
300 Willow Valley Lakes Drive
Willow Street, PA 17584

Acknowledgement

I wish to acknowledge the help of my friend and neighbor, Lewis Anderson, who helped me with the conversion of my final form of this booklet into electronic form.

Table of Contents

Tornado Introduction

I grew up in Kansas and lived many years in Oklahoma, Tornado Alley. I was always interested in tornadoes. I have collected information all my life. Now with the news about the F5 tornado that hit Tuscaloosa, Alabama, in April 2011, I have decided to put my thoughts and ideas on paper.

I did not realize just how bad this year had been for tornados until the CBS "Sunday Morning" TV program on September 4, 2011, announced that there had been "544 deaths in 21 states" from tornados this year. I understand that 160 people were killed in the 2011 Joplin, Missouri, tornado.

What is a Tornado

A tornado is a mature cumulonimbus cloud. Base is about 500 feet above ground; top of the cloud is about 50,000 feet. Upper atmosphere winds blow the top off the cloud forming an anvil-shaped cloud. Vertical circular winds inside the cloud look like an elongated football. Rain falls. The wind picks it up and takes it high in the cloud. Then rain falls again. If you have ever been in a rainstorm where the drops of water are very big, you have probably been under a cumulonimbus cloud.

A resident of Moore, Oklahoma, standing outside his storm shelter saw rain coming down and going back up into the cloud.

It can be very cold in the top of the cloud. Rain being sucked back up into the top of the cloud can become hail. The hail falls, The vertical wind takes the hail back up in the cloud and the hail stones get bigger and bigger depending on how many times the hail is pulled backup into the top of the cloud.

So in this cumulonimbus cloud we have a vertical wind going round and round. Now if you approach this cloud at right angles with an external wind, the internal wind which I have shown in the sketches as an elongated football, will start to spin. Weather researchers are not completely certain as to how this happens. But the spinning vertical wind becomes an even more elongated loop and now you have a tornado.

Sketch of a Cumulonimbus Cloud before It Becomes a Tornado.

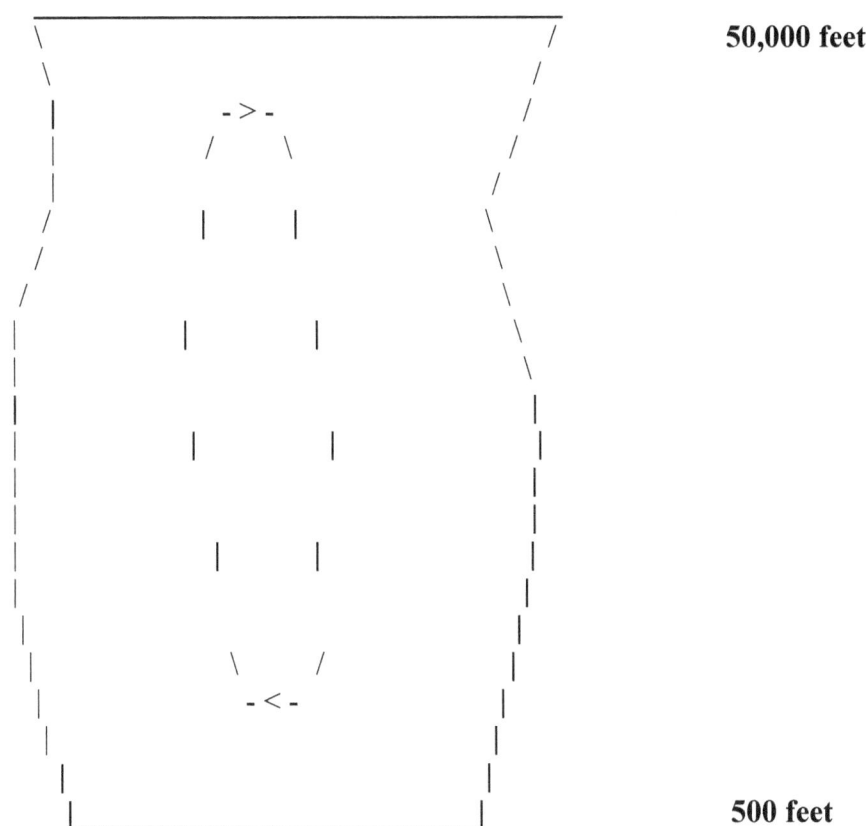

50,000 feet

500 feet

Ground level

A sketch of a Cumulonimbus Cloud showing the vertical internal wind.

Sketch of a Cumulonimbus Cloud after It Becomes a Tornado

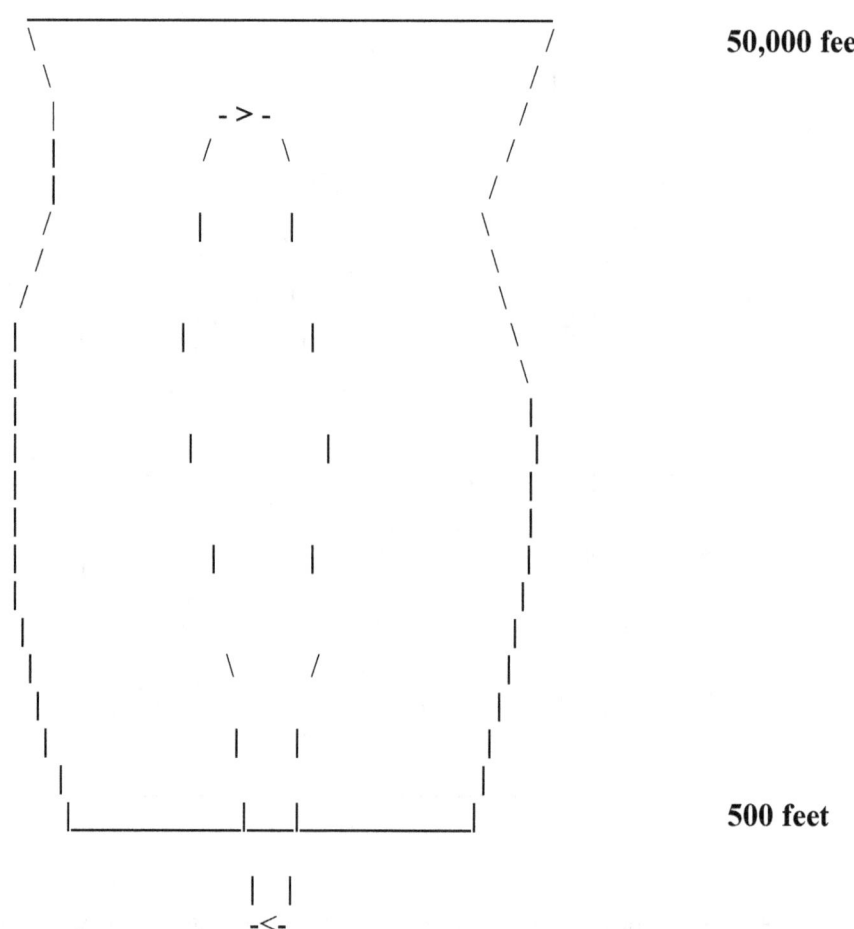

50,000 feet

500 feet

Ground level

A sketch of a Cumulonimbus Cloud showing the vertical internal wind that now extends to the ground and becomes a tornado. Also, there is a circular wind around the axis of the vertical wind which I cannot show in this drawing.

Supercell

In modern technology, a potentially dangerous cumulonimbus cloud is called a "Supercell."

A supercell is defined as a severe thunderstorm that contains rotation. While many ordinary thunderstorms are similar in appearance, supercells are distinguishable by their large scale rotation.

A supercell is a thunderstorm that is characterized by the presence of a mesocyclone; a deep continuously rotating updraft. These storms are sometimes referred to as a rotating thunderstorm. Of the four classifications of thunderstorms (supercell, squall line, multi-cell, and single-cell), supercells are overall least common and have the potential to be the most severe. Supercells are often isolated from other thunderstorms, and can dominate the local climate up to twenty miles away.

Supercells are most common in the Great Plains of the United States.

Tornado Rating Scale

Tornado strength was defined as an F1 (smallest) to F5 (deadliest).

The revised, updated and improved EF (Enhanced Fujita) scale is now used instead of the old F scale. The numbers correspond with the peak 3-second wind gusts as follows:

EF 0 - 65 to 85 MPH

EF 1 - 85 to 110 MPH

EF 2 - 111 to 135 MPH

EF 3 - 136 to 165 MPH

EF 4 - 166 to 200 MPH

EF 5 - over 200 MPH

The tornado has no color until it starts picking things up off the ground. It is usually picking up dirt; so its color is usually black.

Tornados vary in size and intensity. We had one go through Kingfisher, Oklahoma, that was two houses wide. Two houses in each block had their roofs damaged. The tornado that hit Greenberg, Kansas, was very wide and as you know destroyed the whole town.

When a tornado passes overhead it is a sudden very low air pressure. The difference in air pressure is what causes the homes to explode.

These cumulonimbus clouds are very dangerous. But you do not have to worry about them when you are flying in a commercial airliner. Pilots go to great lengths to avoid them.

Hail

Cumulonimbus clouds produce hail. The more times the hail circles inside the cumulonimbus cloud, the bigger becomes the individual hail stones.

I lived in Ardmore, Oklahoma. My wife and family were going out in the country to visit an office mate who had rented a farm. (I was a Gilfillan Radar Field Engineer; he was an Air Force Master Sergeant in charge of all tower and radar operators.) Just as we pulled up in his driveway we realized we were in a hail storm. Hail stones the size of golf balls were hitting the car. The wife was a little shy and reluctant to go to their house. I said, "We don't want to stay in the car. If the hail stones become bigger, the children could be traumatized." So we went to the house with hail stones falling around us yet there was blue sky overhead.

When the storm finally arrived, we were watching it from their porch. The hailstones were the size of tennis balls. They were not just falling from the cloud; they were being thrown down. It did not rain this day; it just hailed. I noted the hail stones were so big they would not roll through the chain link fence.

My car, a 1950 DeSoto, looked like someone had taken a ball-peen hammer to it. For some strange reason, the windshield was not damaged.

The smaller golf-ball-size hail stones that we experienced earlier had been thrown out of the top of the on-coming cloud.

I remember reading about hail stones the size of grapefruit in Louisiana one year.

Hail and Crop Damage

When hail hits a wheat field and the wheat is about ready to be harvested, the hail knocks the grains of wheat out of the head of the wheat. The hail damaged wheat grains are now on the ground and cannot be harvested. Picture yourself; a wheat farmer with a good crop of wheat growing and a hail storm knocks a large portion of your crop to the ground where it cannot be harvested. Thus, many farmers have hail insurance. My high school principal had a summer job of being a hail insurance adjuster. I was always looking for ways to make money in the summers, I checked out becoming a hail adjuster; but college ran into mid June which was too late to start working as a hail adjuster. The start of doing a hail adjustment is to go into a wheat field damaged by hail, cut off 100 stalks of wheat; count the grains of wheat left; count the empty hulls where the grains of wheat had been . From this the adjuster determines the per cent of damage.

Disarming a Cumulonimbus Cloud Before It Becomes A Tornado

Hail damage to American grain farmers is costly and in America is taken for granted. This is not so in Russia.

In America, the cumulonimbus clouds are formed and build in the lee of the Rocky Mountains; then they march across Kansas and Oklahoma.

They have a similar situation in Russia. The cumulonimbus clouds build in the lee of the Ural Mountains.

In grade school we always read the "Readers Digest". I distinctly remember reading how the Russians had solved the hail problem. They used antiaircraft guns to fire a shell of dry ice crystals into the cumulonimbus cloud to dump it early before it matured into a cumulonimbus cloud loaded with hail or a tornado. Rain now was much better than hail later. In recent years, I have tried to confirm this story by checking the internet; but, to no avail.

Seeding a water-laden cloud with dry ice crystals will dump the cloud and make it rain. I remember reading in "Look" magazine about the rancher in New Mexico who would take off in his private open cockpit airplane with a bucket of dry ice. He would fly above the clouds and with a heavy glove he tossed out handfuls of dry ice crystals. He ALWAYS LANDED IN THE RAIN.

Making it rain on your own property is acceptable to all in our society.

Making it rain in certain parts of the country, especially the Bible Belt can be a problem: "Who authorized you to be God?" After the damage in Alabama, North Carolina, and Missouri in 2011, perhaps this attitude might have changed.

My thoughts in 2013, cloud seeding would be an easy job for a UAV (Unmanned Aviation Vehicle or commonly referred to as a Drone).

Minimizing the Damage to Buildings

When a tornado passes overhead it is a sudden very low air pressure. The difference in air pressure is what causes the homes to explode. Usually, the building explodes as if a bomb had been planted inside the building. Reinforced-concrete buildings may survive but may have many windows blown out.

We had a very old well-built house in Silver Lake, Kansas, that was hit by a tornado. When the house exploded, the walls did not just pop out. The floor separated. Part of the floor went out with one wall; the other part went out with the opposite wall. The residents were using the basement for a storm shelter. When the house exploded, the floor separated and dumped all the kitchen appliances on the residents in the basement. A four year old boy was hurt badly.

I saw a new modern house that had been a home with a brick exterior. The tornado sucked the brick off the one side of the house strewing them out into the lawn.

When I lived in Norman, Oklahoma, I would go around the house opening windows where ever I could that it would not rain in. As the storm moves, the wind direction changes. I would keep as many windows open as I could to allow the pressure to equalize if the tornado was very close.

The difference in air pressure may cause the roof to be lifted off the building.

In a 2011 tornado, one building looked very normal from the outside; but, the inside ceiling and wiring were a mess. The tornado had picked the roof up; the pressure equalized quickly and the tornado dropped the roof back down.

My idea is to have several swing-out panels in a building that would let the air pressure equalize quickly thus protecting the building or home from permanent damage.

With a swing-out panel for tornado air pressure equalization, the building would be too easy for burglars to lift the swing-out panels and enter a building. To counteract this, I would suggest an iron grill or web inside the building.

I visualize swing-out panels hinged from the top. These would swing out when a tornado passes over but swing back closed to keep the rain out. Meanwhile, the iron grill inside the building keeps the burglars out but would not impede sudden pressure differences that actuate the swing-out panels.

Wind Damage

Besides potential damage from pressure difference, there is high wind damage. In an F5 tornado the high wind alone can do considerable damage. I am not sure if they even know high the wind speed is inside an F5 tornado. Any piece of debris or any object not tied down can become a very destructive missile in a high wind.

The swing-out panels might protect your building from an F1, F2, or F3 tornado. I have a hunch that nothing can protect your building from an F-5 tornado. I picture an F-5 tornado as whirling, swirling debris that causes damage and adds to the whirling, swirling debris which creates even more damage until the area looks like it had been run through by a giant blender.

I remember when I was in the US Navy in Corpus Christi, Texas, in 1954. We were under Hurricane Condition 4 all summer long every summer. Everything had to be tied down. A blowing bucket in a hurricane wind becomes a deadly missile.

If I were building a home in Tornado Alley today, I would have old-fashion shutters for the windows. They would be spring-loaded so that when tripped, they would close and lock shut. I would have a simple anemometer, built into my weather vane, rigged to trip the shutters when the wind reached a certain speed.

The horrible irony is that in 2011, tornados are everywhere, not just in Tornado Alley.

Lightning Damage

There is a great discussion about lightning in Wikipedia. I recommend it as a great start for learning about lightning.

I have seen a lot of lightning. However, I was not aware of how destructive it could be until I heard my neighbor's story. So, I talked to them extensively and took notes. Then I wrote it up and showed my work to them to make certain that I had their story correct.

Their story follows:

"Background information on the barn:"

The bank barn was 65 feet by 100 feet with a metal roof. An unused 46 foot stave concrete silo stood next to the barn. There was a metal covering on the silo.

For lightning protection, the barn had 3 lightning rods, each about 2 ½ feet high. The lightning rods, the barns metal roof, and the silos metal cover were all properly grounded with 5 foot grounding stakes.

The bottom floor was open and sometimes used for shelter by farm animals. On this night, there were no animals in the barn.

On the next level up were 100 ton of baled hay, 50 ton of baled straw, tractor, and a baler. A car was parked on this level just outside the barn and backed up towards the barn. Also, there was a stack of milled lumber that was all cut, tongue and grooved, ready to be used to wainscot a room in the home.

Along side the barn was an elevated 250 gallon gasoline storage tank that had just been topped off (tank was full).

The weather that night: Heavy rain in sheets that lasted for hours, and plenty of lightning.

The owner was standing in his home on the porch looking at the lightning. He was standing about 120 feet from the barn.

Suddenly, a very large bolt of lightning came down from about a 60 degree angle. The owner said to himself, "This looks like it could kill."

The lightning bolt hit the silos metal cover, exploded very loud and destroyed the cover.

The lightning then came into the barn under the edge of the metal roof.

The lightning then exploded inside the barn; blew out both ends of the barn and set the barn on fire.

The owner immediately ran to move the car. He started the car (keys were always left in the car) and drove it out in to a field. The tail light lenses were already starting to melt from the heat of the fire.

The owner went back to the barn and tried to start the tractor. Two tries and it wouldn't start. So, he abandoned the tractor and quickly left the burning barn.

Meanwhile, the fire truck arrived from 2 miles away. The fire chief said, "There is nothing we can do to save this barn." Three firemen kept one fire-hose on the 250 gallon gasoline storage tank. (The next day the gasoline storage tank was topped off again. It took 75 gallons of gasoline. This is what evaporated during the fire.)

The barn burned to the ground. The unused silo still stands but no roof on it.

Meanwhile, the super heavy rain continued through the night. The metal roof stayed in position and thus the rain did not provide any assistance to the firemen."

My hunch is that the rain and the lightning came from a very large cumulonimbus cloud (a supercell).

So, I think that controlling the size of supercells by dumping them early, could possibly diminish the severity of lightning.

Old and New Theories

The old scientific theory was that there was a pressure differential that caused the homes to explode from a tornado. The new scientific theory is that it is the wind swirling though the house after a window or door is broken down that causes the damage that looks like an explosion. My personal thoughts are that I don't think the experts really know. Has anyone ever had instrumentation set up to measure what goes on during a tornado? I do not think so. Also, the data and the damage can vary according to whether it is an F1, F2, F3, F4, or F5 plus the width of the funnel.

Someone needs to design a black box to record important data. We have black boxes in airplanes so that the experts can study the cause of an airliner crash. We need something like this for tornados. Then, we could build buildings that would be safer.

Tornado Shelters

Every homeowner needs to think about a tornado shelter for his or her family.

You need to have one big enough to have swing down simple bunks. In Oklahoma, there is an average of 2 tornados on the ground every day in the month of May. And, as my co-worker pointed out to me, you need to get a good night sleep as you have to go to work the next day.

If the shelter is in the basement, it should be in a separate room with cinder block walls and a separate roof over these walls. The roof should not be connected to the first floor of the house.

When I lived in Norman, Oklahoma, in the 1960's, all the homes in our development were built on a slab with no basement. Some of my neighbors were building tornado shelters under their driveways with an entry door to their shelter in the garage.

Kevlar® Shelters

DuPont makes Kevlar®. It is used for making bullet-proof clothing for our troops. It is also used to make tornado/hurricane shelters. These shelters range in size from four feet square to twelve feet square with eight foot ceilings. They are bolted to the floor.

In 2011, a woman and her cat survived the Joplin, Missouri, tornado by going into her 4 x 4 Kevlar® tornado shelter that was bolted to her garage floor. She and her cat were together in their Kevlar® shelter for about ten minutes while the tornado did severe damage to her home.

I went on the internet and googled "kevlar"+"shelters" and found suppliers of Kevlar® tornado shelters.

A Kevlar® tornado shelter might be the most inexpensive tornado shelter.

Tornado Shelters for Schools

I was just ready to go to press with this booklet, when Moore, Oklahoma, got hit with a tornado in May 2013. Two schools were hit, destroyed, and some children were killed. Prior to this event, the theory was that a concreted-steel-reinforced building was a safe place to be during a tornado.

Schools need a tornado shelter. My suggestion is as follows:

Rather than putting a shelter in the school's basement, I would recommend an underground shelter under the outdoor playground.

The shelter should have at least two connecting tunnels to the basement of the school. The tunnels should be wide enough so that the student body could be moved quickly from the school to the shelter.

The shelter might also be used as a community shelter on weekends and summers when school is not it session.

In a severe tornado the school might be so damaged that the exit from the shelter after a tornado may be blocked by debris. Therefore, the shelter will need at least two exit doors on to the playground. These doors could be entrance doors for the community when school is not in session. These doors need to be as far away as possible from the school to minimize debris from the school being dumped on the exit doors. The doors need to be heavy duty.

A small gasoline generator needs to be available to power the lights in the shelter and also to provide a small amount fresh air being blown into the shelter. Fresh air input openings need to be rugged construction such that a tornado cannot vacuum children out of the shelter's damaged air input openings.

There needs to be a system so that the emergency generator can be turned-on from any entrance door.

To prevent any improper use of the shelter, the local police need an indicator of use.

This could be a summer project for a school.

If any school decides to do this, please let me know. If there are future editions of this booklet, I will list the schools and a contact person's Name, Address, Phone No., and E-Mail.

Tornado Stories

As a substitute teacher, I occasionally have an opportunity to tell my tornado stories when I am teaching science. Here are some of my stories not already mentioned:

I was working on a climate research project for a professor at the University of Oklahoma. I remember reading this story in one of the journals:

An Air Force pilot flying a weather research plane spotted two lines of thunderstorms on his radar. He decided to fly down between the two lines taking research readings on the way.

Just as he entered this 'alley-way' between the two lines, the end thunderstorm in the second line moved up closing off the exit path planned by the pilot. The two lines of thunderstorms were now formed into a narrow 'U'. There was no easy route for the pilot; he was in the 'U'. It was too narrow for him to turn around and exit. It did not look good for this pilot and the big weather research plan.

He made it out OK. He picked a line between two thunder-storms. It was a very rough flight; but, he made it out courtesy of his flying skill and good luck. His plane could have been easily destroyed by the winds in a thunder-storm.

A Greyhound bus driver going through a small town in Oklahoma noticed a tornado coming head on. The driver decided to go around the block. The suction caused by the bus caused the tornado to follow the bus. No one was hurt; but, it was a close scare.

When I was in the US Navy, I remember reading about a Navy pilot. He had a problem in his jet. He bailed out and when his parachute opened up, he was inside a cumulonimbus cloud. He went round and round, up and down inside this cloud for over 30 minutes. His parachute was wrapped around him. Suddenly, the cloud tossed the pilot out; the parachute became untangled and lowered the pilot safely to the ground.

A boy was caught in a field when a tornado came through. He lay on the ground hanging on to grass to keep from being picked up. He survived. Later, he became a Professor of Electrical Engineering at Kansas State University. There are high levels of lightning with a cumulonimbus cloud. He developed a tornado warning device for homes. He used a modified AM radio circuit. 120 lightning strikes per second would trip the home alarm.

(Note: Lightning is what causes most of the static on an AM radio. That is why we usually listen to FM radio which is static-free.)

In June 2011 a 13 year old boy was killed by lightning. The farm was not too far from where I live. He and his father were baling hay. The father was driving horses pulling the baler.
The boy was walking alongside the bailer. The father felt his hair stand up straight on the back of his neck; then heard the loud clap of thunder; he looked back and his 13 year old son lay dead on the ground. The medical authorities could not revive him and pronounced him dead from a lightning strike.

I remember in Kansas, cows tended to congregate under a tree during a heavy rain storm. To protect the cows from lightning, the farmers would put a lightning rod in the cow's favorite trees.

Lightning causes thunder. The image of lightning travels at the speed of light (186,000 miles per hour). The sound of thunder travels at the speed of sound (550 miles per hour). So for every mile from lightning, about 5 seconds elapses between seeing and hearing it. As a young lad in Kansas, I could make a quick calculation of the distance to the storm. I would see the lightning and start counting the seconds until hearing thunder; 1, 1000, 2, 1000, 3, 1000, etc. On many occasions the time between seeing the lightning and hearing the thunder was 5 seconds so the lightning was about a mile away.

Growing up in Kansas and living in Oklahoma, I was in Tornado Alley. However, I never saw a cumulonimbus cloud there. It was always raining too hard. The one that hit us with tennis-ball-size-hail in Oklahoma was not visible because of the haze. I finally did see a cumulonimbus cloud in Arizona; a beautiful large white anvil marching across the sky.

On June 3, 2011, a lady was driving in Southwestern Kansas. There were lightning strikes headed to the ground from the South and West! The wind picked up and was rocking her Suburban as she drove. Orange highway cones were blowing across the road. She dodged them knowing that the cars behind her wouldn't make it through the road construction because the highway was littered with cones and signs. She saw a big orange highway road closed sign spinning around along with dust, leaves, and debris. The spinning sign was probably 4 feet by 8 feet plywood lumber. It was painted orange with "Road Closed" written on it. The event seemed to be only 10 to 20 seconds long. She just held her breath and headed toward the underpass. She recited Psalm 91, from the Bible, which is a prayer of protection. As soon as she reached the underpass, all the blowing wind and debris stopped as if nothing had ever happened

She was very blessed and received no damage. One side of her car looked like it had been in a mud ball fight and was plastered with chunks of mud.

Summary

Remember this:

    Mother Nature can be very gentle or an indiscriminate killer.

So, as I tell my students:

    "PAY ATTENTION."

Notes

My Personal Tornado Protection 'To-Do List'

Our Family's Personal Tornado Survival Plan